ISBN: 978-0-6455554-0-0

General enquiries
Info@1000tales.org.au

Shop
www.1000tales.org.au

First Printed in 2020.
©Copyright 1000 Tales.

All rights reserved.
This book or any portion thereof may not be reproduced, stored in or introduced in a retrieval system, or transmitted, in any form or by any means without the express written permission of 1000 Tales Co-op Ltd except for the use of brief quotations in a book review.

Written by
Ameera Karimshah

Illustrated by
Atiya Karimshah

Design Consultant
Taarun Krushanth

FOR TAARUN AND HARUN.
THE HEART AND SOUL.

MAY HISTORY RECOGNISE YOUR
ACHIEVEMENTS AND THOSE OF OTHERS
LIKE YOU.

WE WOULD LIKE TO ACKNOWLEDGE THE TRADITIONAL CUSTODIANS OF THE CONTINENT OF AUSTRALIA. WHOSE CULTURES ARE AMOUNG THE OLDEST LIVING CULTURES IN HUMAN HISTORY AND WHOSE LANGUAGES AND KNOWLEDGE HAVE INFUSED AND INHABITED THIS LAND FOR MILLENNIA.

WE RECOGNISE THEIR CONTINUING CONNECTION TO THE LAND, WATERS AND CULTURE AND WE PAY OUR RESPECTS TO THEIR ELDERS PAST, PRESENT AND EMERGING.

If you love learning about science then you are probably familiar with the achievements of the ancient Greeks like Pythagoras and Galileo but did you know that the Greeks did not do it on their own.

In fact in some areas the Greeks lagged behind the rest of the world by hundreds sometimes thousands of years (mathematics, astronomy etc.) The Greeks travelled, learned, borrowed from and collaborated with people all across Asia, Africa and the middle east.

This collaboration, didn't start or end with the Greeks, it went on long before and after them. Although the Greeks made many advances, much of what they left behind was also problematic.

This is the story of how one of those problems was solved.

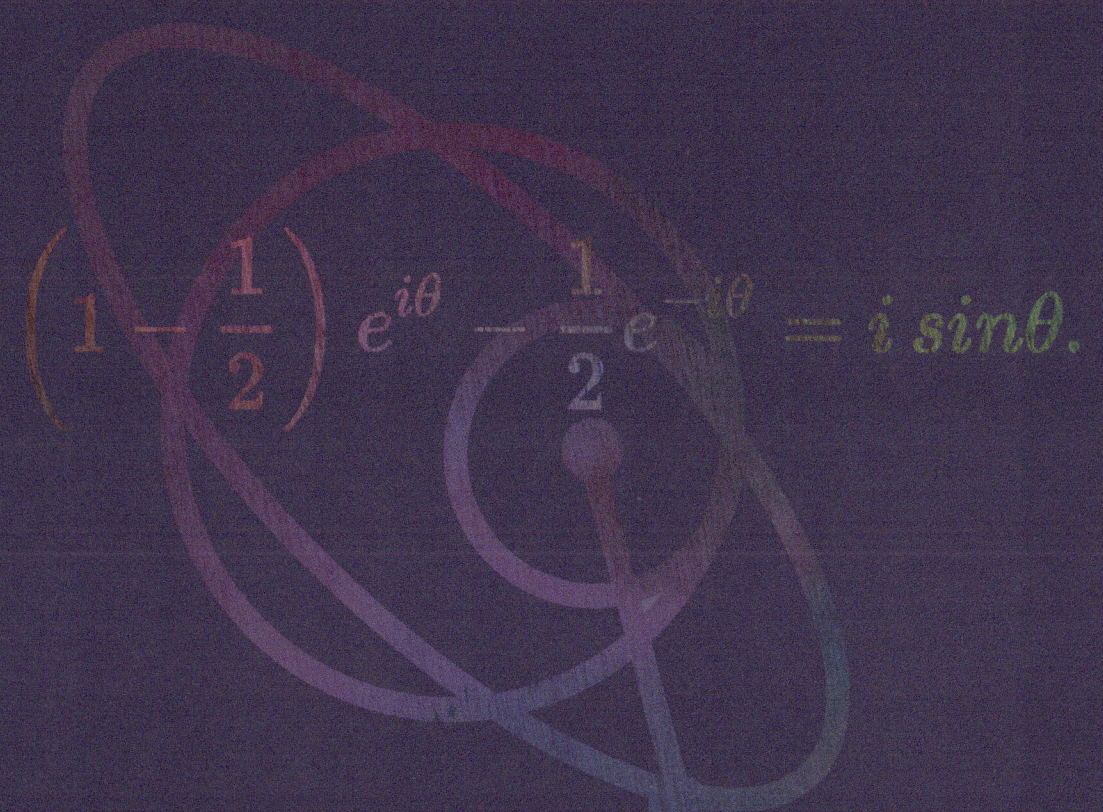

$$\left(1 - \frac{1}{2}\right) e^{i\theta} - \frac{1}{2} e^{-i\theta} = i\sin\theta.$$

AL-TUSI AND THE EQUANT PROBLEM

SECRET
SECRET
SCIENCE
SCIENCE

DID YOU KNOW?

The ancient Greeks believed that the earth was the centre of our planetary system and everything in the sky revolved around us!

This idea was put forward by a famous philosopher named Aristotle and 'proven' mathematically by an astronomer named Ptolemy in a book called the Amalgest.

It was a difficult idea to prove because while traveling across the night sky some planets appear to suddenly change direction and start going the other way! So how could they be moving around the earth? In order to explain this, Ptolemy came up with a very long and confusing theory, which turned out to be physically impossible. This is known as the 'equant problem'.

TODAY WE KNOW THAT THIS HAPPENS BECAUSE THE EARTH OUTSPEEDS OTHER PLANETS, LIKE MARS FOR EXAMPLE, AS BOTH PLANETS ORBIT THE SUN. KIND OF LIKE ONE CAR PASSING ANOTHER.

Over a thousand years later - in 1543, Polish astronomer Nicolaus Copernicus solved the 'equant problem' in a book called 'Concerning the Revolutions of the Heavenly Spheres.'

In it was mathematical proof that the sun was the centre of the solar system and not the Earth. His work earned him the title of 'The Father of Modern Astronomy.'

Nicolaus Copernicus

Born:

19 February 1473 in Thorn, Poland

Died:

24 May 1543 (aged 70) in Frauenburg, Poland

Known for:

Astronomy, Canon law, Economics, Mathematics, Medicine, Politics

FOR HUNDREDS OF YEARS PEOPLE THOUGHT THAT COPERNICUS HAD INVENTED THIS MATHS BY IMPROVING ON PTOLEMY'S WORK - THE AMALGEST.

However, In the 1950s a set of documents were found at the University of Beirut in Lebanon that contained geometry identical to Copernicus's model only these had been written in Arabic and dated all the way back to the 13th Century.

The two formulas that were central to Copernicus's work were originally devised by scholars Nasir al-Din al-Tusi and Mu'ayyad al-Din al-'Urdi.

Nasir al-Din al-Tusi was a Persian mathematician who in 1247 wrote a paper titled Tahrir al-Majisti (Commentary on the Almagest) in which he solved the centuries old problem that had stumped Ptolemy and other ancient Greeks.

Mu'ayyad al-din al-'Urdi was a Syrian Astronomer who worked closely with al-Tusi. It appears that Copernicus simply copied al-Tusi's and al-'Urdi's formulas.

HOW DO WE KNOW?

Well, Copernicus's math contains random details that are identical to al-Tusi's.

For example, the labels given to specific geometric points.

These are usually chosen at random by the mathematician.

It can't be a coincidence that the geometric labels used by Copernicus were almost exact translations of al-Tusi's.

FOR EXAMPLE,

the point labeled with the Arabic

letter alif أ by al-Tusi was marked

 by Copernicus.

The Arabic ba ب

was marked  and so on.

Each Copernican label the phonetic equivalent of the Arabic — almost all were identical!

There was only one exception.

The point marked as  by copernicus was

zaa ز in al-Tusi's diagram.

In Arabic script, however, a handwritten

zaa ز .ould be easily mistaken

for a faa ف

Something else that is quite unusual is that Copernicus did not include all the proofs in his work, which is almost unheard of for such a detailed mathematical model.

As it turns out the missing proofs had already been published 300 years earlier by al-'Urdi.

Copernicus might have come into contact with al-Tusi's work while studing in Padua Spain, which had previously been part of the Islamic empire.

There was a cache of ancient Arabian research documents stored at the University of Padua that Copernicus would probably have had access to.

$$\left(1 - \frac{1}{2}\right) e^{i\theta} - \frac{1}{2} e^{-i\theta} = i \sin\theta.$$

It is difficult to know for sure but history tells us that Europe and the Middle East don't always get along and perhaps Copernicus didn't ireference al-Tusi's and al-'Urdi because he thought it would make people less willing to accept the ideas he was presenting.

Even today, some people find it difficult to believe that the idea didn't come from Copernicus. They say that Copernicus came up with the mathematical equations all by himself. After all, they are logical solutions to a very old problem.

A lot has been written about **NICOLAUS COPERNICUS** and there is no doubt that he was a great scientist.

But, what do we know about

NASIR AL-DIN AL-TUSI

and

MU'AYYAD AL-DIN AL-'URDI.

Ibn Muhammad Ibn al-Hasan al-Tūsī

Known as

Nasir al-Din al-Tusi

Born

18 February 1201 in Tus, Iran

Died

26 June 1274

Known For

Mathematics, architecture, philosophy, medicine, science and theology.

Moayad Al-Din Al-Urdi Al-Amiri Al-Dimashqi

Born

1200 in Nisba al 'Urdi, Syria

Died

1266

Known For

Mathematics, architecture, philosophy, medicine, science and theology.

Known as

Mu'yyad al-din al-'Urdi

Nasir al-Din al-Tusi was born in the northeastern part of Iran in the valley of the Kashaf River.

At the time, the enormous armies of the Mongolian empire were spreading across Asia and Europe.

In particular the Mongolians targeted Muslims like al-Tusi and his family because they believed that the Muslim people posed a threat to their goals of being the most powerful empire in the world.

This meant that Al-Tusi's early life was very difficult and he had very little peace to pursue scientific work.

Luckily al-Tusi was born into a family that put a high value on knowledge.

So, in spite of the constant threat of war al-Tusi's father made sure that he attended school where he learned mathematics - particularly algebra and geometry.

Outside of school his uncle became his first mentor and sparked a lifelong interest in learning with lessons on philosophy, logic and physics.

As a teenager al-Tusi moved to Nishapur to continue his studies in philosophy, medicine and mathematics. While in Nishapur he gained a reputation for being very very smart.

The war, however was not going away. So in his early 20's al-Tusi joined a group of soldiers fighting against the Mongolian army and took refuge in the castle of Alamut in the Elburz Mountains.

It was an impenetrable fortress, which meant that while training with the other soldiers al-Tusi was safe from the daily struggles of the war.

This isolation gave him the space to complete some of his best work. It was here that he wrote a number of important books on logic, philosophy, mathematics, and astronomy, including the Tahrir al-Majisti from which Copernicus might have borrowed his formula.

In 1256 the castle of Alamut was invaded by the Mongolian army. Luckily, the attack was led by a General named Hulegu, who had a great interest in science and chose to preserve al-Tusi's work while destroying many other things. Hulegu was so impressed by al-Tusi that after spending some time as a prisoner he was appointed as the General's scientific advisor.

This was a great opportunity for al-Tusi as it finally gave him the resources to fully pursue his work.

Soon after his appointment, in 1259 al-Tusi requested the construction of a great **OBSERVATORY**

in a place called **MARAGHEH**

which was the capital of Hulegu's territory.

He invited al-'Urdi who was in Damascus at the time, to help establish it as the centre of Astronomy for the Middle East and Asia, with scholars visiting from as far away as China.

The ruins of this great Observatory are preserved under a brass dome just west of the city today and will eventually be turned into a museum to preserve al-Tusi's legacy.

Al-Tusi's work included accurate tables of planetary movements and a star catalogue. He invented a number of different astronomical instruments and vigorously pursued knowledge in science, mathematics and philosophy.

He established trigonometry as a mathematical discipline independent of astronomy and worked with scholars from across the region to grow our knowledge of the universe.

Importantly, he wrote commentaries on many Greek texts, disproving ideas and adding to a growing body of work that is not often acknowledged in modern accounts of science history.

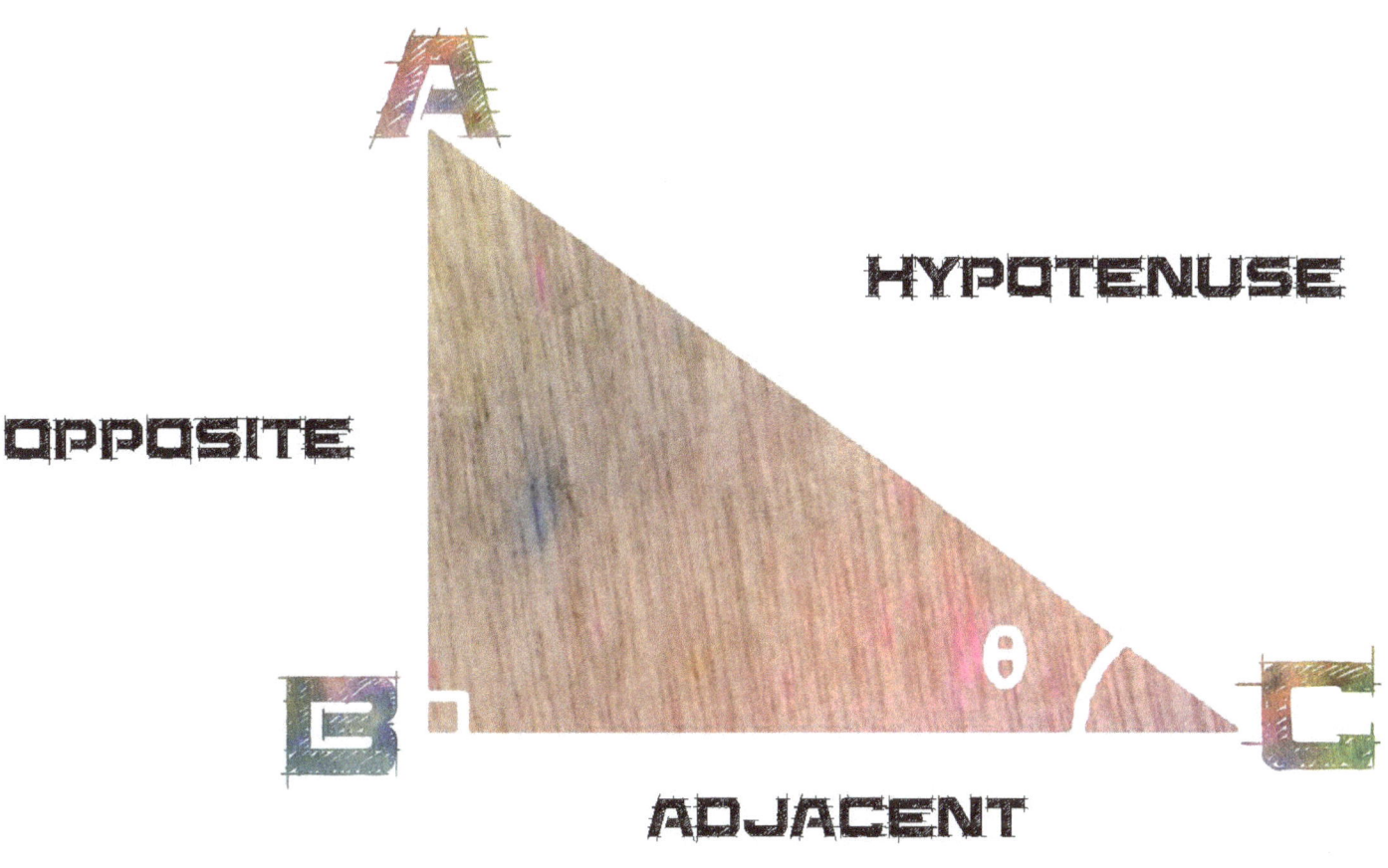

After a lifetime of great scientific achievements al-Tusi died in 1274 and left behind many students including al-'Urdi's son who continued his legacy and would go on to make their own scientific discoveries.

FOR MORE SECRET SCIENCE FOLLOW US:

https://www.facebook.com/thousandtales

@1000tales

www.ingramcontent.com/pod-product-compliance
Lightning Source LLC
Chambersburg PA
CBHW041711290426
44109CB00028B/2849